Deadly Disease

Ashlee Wilson

Chapter One

It was a normal day, just like every morning, Katie was heading into her office to work on her research. Ever since she lost her mom three years ago, she made it her life goal to find a cure for not only cancer, but for all deadly diseases. Katie didn't want others to go through the pain her mother went through, especially near the end, when her mother couldn't even get out of bed. Katie spent hours testing different blends of medicine, trying to find the perfect cure. She just started testing on animals a little over two months ago. So far Katie has not had a break through, but she is not going to give up until she does. Like most nights Katie stayed late trying different medicine on each animal. Getting frustrated, Katie tried the last blend of medicine she had for the night, she locked up the animals and decided to go home. At about 5:00 in the morning she gets a phone call saying she needs to head into the office right away. Not knowing what was going on, she got up and got dressed, then drove to the office where she was met by Eric the night security guard. In a panic he quickly took Katie up to the third floor where they kept the animals. When Katie entered the room she saw Molly, one of the monkeys she was testing on that night. Molly was going crazy bouncing off the cage. Katie didn't know what was going on with Molly, so she decided to sedate her until she could figure out what was happening. Right away, Katie went up to her office, she started going through all her notes to see what was given to Molly. After looking at all her notes, she couldn't find anything that should make Molly act the way she was. Katie ran all kinds of test on Molly, but the test kept coming back clear. Nothing was giving her answers as to why she was behaving like she was. Katie went over all the side effects of the blend of medicine that was given to Molly, and still she had no answers. She has not seen anything like this, but she has only been doing this for a little over two months. She decided she could use some help on this, so she decided to call her friend Sarah, who has been doing this for two years. After talking to Katie, Sarah decided to go down to help Katie out. She told her that it may be late when she arrived, because she had some work she had to finish up before she could leave. When Sarah finished up what she was working on, she packed up a few things she thought would help

Katie's situation. Katie told Eric, that she had a friend that was coming down to help her, and when she arrived to send her up to her office. On her way up she decided to check in on Molly, who was resting comfortable. Her heart rate was steady, but her blood pressure was high, so she decided to bring her back up to her lab where she was doing her research the night before. Molly was living her life in an animal hospital not expecting to live. The doctors that were taking care of Molly found brain cancer. Katie could see the pain Molly was in and explained to the doctors what she wanted to do for Molly. They agreed to let Katie take Molly with her and now it was breaking Katie's heart to see her this way. When Sarah arrived, Eric told her that Katie was waiting in her office for her and sent her right up. When she got up to Katie's office, she found Katie at her desk, papers all over the place. She could see that Katie was very frustrated so she asked Katie if she needed to take a break before they got started. Katie told her she was fine, that she just wanted to find out what was happening to Molly. Sarah hugged Katie and told her that they would find out what was happening to Molly, then asked for Katie's notes so they could get started. Katie told her about all the test she ran, and gave her the results for her to look over, making sure she didn't overlook anything. Sarah studied the test for a while, and could tell why Katie was so frustrated, then looked at Katie and asked "did you think to do a CT Scan on her"? Katie told her she ran one when she brought Molly here, but hasn't ran one since. They hooked Molly up and started the CT Scan, and it showed that since Katie started Molly on the medicine that the cancer had grown. Sarah and Katie talked, and decided that since Sarah had a larger lab and more resources, that Molly should be treated in her lab. They got Molly ready for the trip, and Katie made sure Sarah had all the notes she needed, before she and Molly headed back to her lab. Katie decided that since she has been working so close with Molly, that she may need to come as well so Molly wouldn't be as scared of her new surroundings. Sarah told her that was a good idea, that they could work side by side like they used to, and could help molly together. Katie told her staff of her leaving and told them she wasn't sure when she would be back, but would keep everyone posted about Molly. When they got Molly settled in, they went up to Sarah's office and started to do some research. They worked all through the night taking notes and running more test. After Molly spiked a fever, Sarah thought they should start her on some strong antibiotics and fluids, because she was afraid that Molly could start having seizures, due to the high fever. They decided that someone should be with Molly at all times, but with them both doing research trying to find out what was going on with her, they didn't have the time to set with her. Both Sarah and Katie wanted to be the ones who stayed and took care of Molly, but talked it out and decided it would be best if they could get a nurse to stay with Molly, then they could do their research and Molly would have the care she needed. Sarah assigned her top nurse, Kelly, at the company to take care of Molly. When they explained what Molly was taking and what she would need to do if she woke up, but told her she should be out for a while. When they thought Kelly was all set, they decided to take a break and go get a bite to get and a few hours' sleep. They told Kelly that if something were to happen to call and they would come back to help. Kelly told them she would be fine, but if something happen she would call. Sarah and Katie checked in on Molly before heading out to get a quick bite, but before they got to get a few hours of rest the phone rang. It was Sarah's secretary Abby, she told Sarah that while she was getting ready to get started working for the day that she heard a loud noise coming from Molly's room, but when she went to check to see what it was, she saw Kelly passed out on the floor. Sarah hung up the phone and told Katie what Abby had found, then rushed right over. BY this time, it was already nine in the morning, Sarah and Katie have not had any rest since two nights ago before this started. When they arrived, Molly was still sedated and Kelly was unconscious. They decided to run some test to see if there was any cause to why Kelly was now unconscious with a high fever. Nothing on the test were showing up anything that would explain why this was happening to Kelly. They ran more test, did blood work, but still nothing. They decided that they should start Kelly on fluids and antibiotics to keep her from a possible seizure. With all the test they ran coming back clear, they seemed to be more lost than before. They headed back up to the office hoping they missed something because of the lack of sleep they had. What could be the cause of all this, the two thought, and why is nothing showing up. They knew they had a long night ahead of them, with both Molly and Kelly with the same symptoms. They knew they would have to treat them differently, but with the both of them only experienced in animals they were lost on what to do with Kelly. They knew they would have to find someone who could give Kelly the care she needed, but knew they had to keep quite about what happen while she was in the care of Molly. Both frustrated and confused decided that they had to find someone they could trust, but who?

Chapter Two

Since they didn't know what caused Kelly to fall unconscious, they decided to take precaution when working with Molly. Katie told Sarah that they needed to set up a meeting with the people they thought they could trust, and tell them what was going on, and what they were doing to prevent this from happening again. So Katie, along with Sarah, called a few people, that they thought they could trust, to come help them out until they found out what was going on. Katie decided to call her top nurse, Adam to come help them. Katie made sure that he knew what was going on, and that they were trying to do all they could to make sure it didn't happen again. Adam agreed, and told Katie he would just tell people he was going out of town for a few weeks, then he would head down to help her out any way he could. While they were waiting for Adam they decided to do more blood work on Molly and Kelly, to see if they may be any changes. The test results were the same as before and there still didn't seem like much hope for anything to show up. Weeks have gone by and they are not anywhere near close to finding out what is going on. Sarah has Emily come to take care of Kelly, but tells her what all has been going on. Emily agreed, and told Sarah that she would help anyway she could. Not even an hour went by, before Abby was running in to tell Sarah that she was walking by Kelly's room, when she saw Emily laying on the floor. Confused Sarah got up and went to check and Emily was unconscious on the floor. Sarah couldn't understand how Emily fell ill so fast, but Kelly was around Molly for several days before she fell ill. Abby started freaking out, she told Sarah and Katie that she didn't know what was going on, but she no longer felt safe being here and she wanted to go home. Sarah told Abby that if she felt safer at home, that she was more than welcome to leave, but asked if she could get her a few things and put in her office. Since Emily was taken care of Kelly and fell ill so fast, she thought someone needed to go check on Adam. Katie told Sarah she would go check on him, when she got to Molly's room she could see Adam laying on the floor. Without thinking she rushes in to Adam, but he is unconscious as well. Katie noticed just like Molly and Kelly, that Emily and Adam were running a high fever. She had a few people come help her get Adam and Emily to a room where she could run some test on them. Sarah thinks it may be time for them to lock the lab down so she goes searching for Katie, but she is nowhere to be found. Abby passes Sarah on her way out, so Sarah asked her if she saw Katie anywhere, and that's when Abby told her she saw Katie leaving. Sarah asked Abby if Katie said anything before leaving, and Abby told her she said something about going back home to her office, but that was all she said. Confused to why, Sarah told Abby thank you and to get home safely. Abby hugged her before she left, and told her that she would come back if she was needed. Sarah didn't think too much into Katie leaving, she just thought that maybe she left some notes back home that she thought would be helpful. She didn't know that Katie ran in after Adam, and that she left freaking out. Sarah heads back up to her office to call Katie's office, but is told that Katie called telling them she was heading back, but hasn't shown up yet. So, Sarah calls Abby, and asked if she mind coming back and looking after things so she could go check on Katie. Abby asked what was going on, and that's when Sarah told her about Katie not making it back to her office. Abby told her she would come back, but to put cameras up where she could watch from the office where she felt safe. Sarah told her that she would install some right away and that if she saw anything out of the ordinary to call her. Sarah tells Abby not to enter any of the rooms that if something happens to call right away. She told her to lock all the doors, that she didn't want anyone coming in after she left, and that she put a camera up at each door so she could see if anyone came to the door. As Sarah is heading to her car to go find Kate, she gets a call, its Katie telling her she made it safe and not to come down, that she would be back soon. Sarah thinks nothing of it and tells Katie she's glad she made it and to call if anything changes. Sarah heads back in, where she informs Abby about what was going on, and she could head back home if she wanted to, but Abby decided to stay. Meanwhile Katie is up in her office searching through her notes to see if she can find anything about falling unconscious and suddenly spiking a fever. She decided that she could use Sarah's help, so she called and told her what she was doing, and to see if she could research the sudden spike in fever. Sarah told her, she would see what she could find and she would call when she found something. As soon as they hung up Katie started feeling strange, she started coughing up blood and sweating. She tries to play it off as she is fine so no one would see something was wrong and be alarmed. Eric comes in to see if there is anything she needed. Katie tells him she's fine, but he can see something isn't right, and ask if she's sure. Katie just nods and goes back to work thinking nothing of it. He leaves her to her work, but knows that there is something going on so he makes a phone call. As he hangs up from the call he hears a loud scream coming from Katie's office. He runs up to see if she's alright, but finds Katie on the floor, blood everywhere. He remembered she was working with Sarah, and gives her a call to let her know what happen. He told her that he didn't know what was going on, but that someone needed to get to the bottom of this. He went on and told her that he made a phone call, and that she would be getting some visitors soon, but before Sarah could ask any questions he hung up. Sarah was on high alert, because she didn't know who was coming or why they were coming. Sarah heads to her office where Abby is setting, and tells her about the phone call she had with Eric. The two of them stayed in the office and watched the cameras, then two hours later two army men were banging on the door. Sarah told Abby to stay in the office that she was going to see what they wanted. Sarah went down and opened the door and the two men introduced their self. One was Sergeant Scott and the other was Lieutenant Lee. They tell her that they got a phone call from a man that said his boss was working on experiments with animals and that one was sick and brought to this lab. She told them, that Molly was here, and told them what was going, then took them to Molly's room. She also told them, that three people started having the same symptoms, and she had them in separate rooms. She also told them she got a phone call that Katie was sick, but didn't know if she was okay or not. Sergeant Scott told her that they had Katie at their hospital on the base, and that was all he could tell her. He told her that they were taking over and he needed her to hand over her notes, and anything she had that was about the situation. They told her they were taking over her office, but she could stay in the lab, but had to stay out of this office. Lieutenant Lee told her they were putting the lab on lock down, that as of now no one was to leave or come in. He told her to go gather the ones not infected and meet him back here. Sarah went and got Abby, and told her what was going on. She asked her to help get the others and she apologized for keeping her here and getting her involved. Sergeant Scott took Sarah, and the others and put them in a room, so he would know where they were and what they were up to. Sarah kept trying to ask questions to try and find out what they wanted and what was going on, but all she got was we can't tell you at this time. She had no contact with the outside world, and didn't know if Katie was dead or alive. She thought all this was strange, and didn't understand why the army wanted to take over. Did they know more than they were letting on, and were they hiding something. She talked to Abby, when the guard left, and told her how she felt about all this, and wanted to know what Abby thought. Abby told her she didn't understand any of this, that it must be something big for the army to get involved. Sarah told her when she got the chance she would look into it, but for now they just needed to do what they said. Sergeant Scott, went and got Sarah and told her she needed to come with him, that there was something she needed to see. Sarah wasn't sure what she might find, but as soon as she turned the corner she let out a loud gasp.

Chapter Three

Sergeant Scott quickly covered her mouth and told her she had to keep this quite. They went in Molly's room, and not only was Molly gone, but the door to her room was broken down from the inside. "Sergeant Scott", Sarah yelled in panic, "what happened here"? He told Sarah as he was doing his rounds he noticed the door was missing, and went to see if the camera caught anything, but it was blank as if someone turned it off. He then told her he came and got her because she was the only one not infected that had contact with Molly, and he wanted her to work with them to find Molly. Sarah told him, that they kept Molly sedated, so she wouldn't be in pain while they tried to figure out what was happening. She told him that when Katie showed her Molly, that she was already sick, and there was no way she should have the strength to walk, must less break a door down. Sergeant Scott reassured her that he would get his best men to find Molly, and that they would do everything to not harm her. Sarah asked, if she could be part of the search, since she kind of knew Molly's case. He told her that would be fine, but that he wanted her to be paired with his best men just in case Molly was not the same. She agreed, but told him they may want to check on the other patients, to make sure they were still in their room. He agreed, but told her that if they found their rooms empty, he needed her to try and stay calm. She nodded, and they headed to the rooms, where they found them to be empty. Just like Molly's room, they found the door broken down from the inside. As they were headed back to meet up with his crew, he got a call from Lieutenant Lee. He was calling because, he got a call from the base and they said Katie has went missing. Sarah could see the worry on Sergeant Scott's face, so she asked if everything was alright. He just looked at Sarah and told her they had a big problem. He told her what Lieutenant Lee had told him, and that he was worried because all the infected was now missing. They decided, that, they needed to remain calm, because they didn't want the ones not infected to panic. Sarah told Sergeant Scott she would go let the others know, that for right now they would be safer to stay where they were. He told her not to say anything about what has happen, they would tell them when they knew more. Of course when Sarah told them, they all wanted to know why, and what was going on. Sarah told them she didn't know, but when she found out she would come back and let them know. As Sarah was getting ready to head back to let Sergeant Scott know that she told the others, the power went out. Since the door to the room the uninfected were in was power operated, she could no longer keep them safe. Everyone was now screaming in panic, and knew that they were not being told everything. Sarah told them that she needed them to remain calm, that she wasn't for sure what caused the black out, but if they all stuck together they would be fine. With everything she knew, this panic was not helping with the situation. Sergeant Scott told Sarah that they needed to get this under control, that all this panic was doing nothing but making it worse. Abby tells Sarah that she was going home, that she knew something wasn't right when they brought Molly here. Sarah told her that she would be better off to stay here, so there would be a group and she wouldn't be alone if something was to happen. Abby told her that she didn't feel safe, that there was to many secrets and nobody knew what was going on, and she felt like someone should know something. Sergeant Scott told her no one would keep her from leaving, but if she left that she wouldn't be able to come back, because they didn't know what was going on outside the lab. She thought about it for a while and decided she was leaving, because no one seemed to know what was going on, but before she left Sarah told her to please be safe and that she was sorry about everything. After Abby left, Sarah started to panic, she was thinking the worst was about to happen. Lieutenant Lee could tell Sarah was about to lose it and went and got Sergeant Scott. He told him, he could tell Sarah was starting to panic and that they needed her to help make sure everyone was safe. So Sergeant Scott went and talked to Sarah, he told her that he needed her, to try her best, to stay calm and help him get blankets, flashlights, and see if they could find some food. Sarah told him there was a small kitchen on the second floor that may have some food, but didn't know for sure. They divided up in small groups and went and gathered all they could find. There wasn't much in the kitchen, but they got what they could. The groups met back at the stairs once they got what they could, but when they were heading back up to the third floor they could see at the top of the stairs someone or something was blocking the door. As they got closer, Sarah let out a loud gasp. With tears rolling down her face, she cried out, Kelly. Shoot her, they yelled, we don't know what she might do. No, Sarah screamed, don't shoot her, she's just scared. Just then Kelly charged at her. Sergeant Scott pushes Sarah out of the way, Kelly misses and rolls down the stairs. Sarah, panics and runs down after her not thinking what she might do. When she got at the bottom of the stairs she opened the door hoping to see Kelly, but she saw no Kelly just glass everywhere. She knew that they were no longer safe here, but as she turned around she saw they followed her down. She told them that they needed to find a new shelter, that she felt like they were no longer safe here. She told them, there was no time to waste, they needed to take what they had and go now. Everyone looked confused, it was dark out and they didn't know what could be waiting on them. One of them spoke up, it was Chris her top researcher, he asked why did they have to go now. and if they would be safer to wait till morning. Sarah told him that it would be better to go when it was daylight, but since no one knew what was waiting on them in here that they should go now. Sergeant Scott agreed morning would be better, but knew if they stayed here they may not make it till morning. They all finally agreed they should go now, so Sarah handed out flashlights and told everyone to stay close. She was so scared, because if things were this bad in here then there was no telling what was going on out there. She knew she could not freak out, everyone was counting on her to get them to safety. They headed out, Sarah and Sergeant Scott were upfront, while Lieutenant Lee was at the back with the rest of Sergeant Scott's men. Since they wasn't sure what to expect when they went out, they wanted to find somewhere close., but safe just until morning. They came up on a warehouse,but before they went in Sergeant Scott and his men went in to make sure it was safe. After searching they came back and told them it was safe and for them to come inside. As they were getting settled for a long night, a loud noise came from one of the back rooms. I thought you check it, yelled Sarah. With a puzzled look on his face, Sergeant Scott told her they checked every room and they all came up empty. He got a few men to go with him to see what they missed, but when they went to check nothing could prepare them for what they saw.

Chapter Four

When Sarah saw Sergeant Scott, she knew that something wasn't right. She asked if he was okay, but he just kept walking. She knew something was wrong , but wasn't sure if she wanted to know. The look Sergeant Scott had on his face said it all. Sarah didn't know if she should go after him or go see what they found. Just as she was about to go see what was going on, Lieutenant Lee came and told her that they needed to go,now. He told her he had no time to explain, that she needed to get everyone and seek new shelter. Sarah asked no questions, she went and told the others that it wasn't safe here anymore. They walked about five miles when they came up on a store. Sarah thought they should have a look inside, to see if they had food and other items they may need. Sarah asked, Lieutenant Lee if he could have some men have a look to see if it was safe. After he had a look around he came back out and told Sarah it was safe, but he needed her to go in the back room. When she went back there Sergeant Scott was in a corner crying. She went over to ask what was wrong, he broke down and told her this was all her fault. She was confused, and asked him why he was acting this way? He just looked at her and kept repeating, this is all your fault. Sarah ran out of the room, she went to Lieutenant Lee and asked him what was going on with Sergeant Scott. Not knowing what to do, he told her that what happened back at the warehouse really got to Sergeant Scott. She wanted to know what they saw and why it made him act crazy, but all she got was just give him time and he will be fine. Sarah just shook it off and went and helped the others get what they needed so they could find better shelter. She didn't know how long she could just act like nothing was wrong. She knew that she had to stay strong, but as soon as they found a safe place she would have to confront Sergeant Scott. They walked for hours ,before they saw an old building, and Sarah told them this looked like a good spot for the night. After a long search, Lieutenant Lee found a safe room where they could lock themselves in. Thinking they were safe for the night they settled in. Sarah went through what they had, and portioned out the food, just to see how long it would last. Although, they had their selves locked in they still took turns standing guard, just to make sure no one or nothing came through the door.

While Sarah was on watch , there was a loud noise that woke everyone up. Without knowing what or where the noise came from, Sarah thought it would be best for them to wait it out there. Everyone agreed, but at first light Lieutenant Lee and Sergeant Scott went to check it out. Although Sergeant Scott was still upset about what he found at the warehouse he knew he still had to keep everyone safe. Sarah decided to go with them, she told the others to lock the door behind them, and not to open it unless it was them. She told them that if they didn't make it back, to use their best judgment and to leave when the whole group decided to. Sarah was not ready for what she was about to see, she was scared it would be the worst, but it was Abby she had locked herself in a room, and Katie was trying to get in. Sarah knew what needed to be done, but she didn't have the heart to do it. So Lieutenant Lee told her he would just sedate Katie and they could make her comfortable. They placed Katie in a room and locked the door so the others would be safe. While Lieutenant Lee took care of Katie, Sarah helped Abby out of the room and headed back to let everyone know what was going on. On their way back Sarah asked Abby how she ended up here? She told Sarah that she thought she might be need back at the lab and was hoping that after she talked to Sergeant Scott he would let her help.She told her when she was almost back to the lab Katie came running out of a warehouse so she had to find shelter and was chased to a near by store where she thought she lost Katie. She told her as she was catching her breath, she saw them leaving the store and followed them. She told her she waited until they were inside before she went in because she knew it would be safe if they stayed, but before she could catch up to them Katie chased her into the room where she safely locked the door. Sarah told her she was glad she was back, that she missed her and needed her to stay and help her. She told her that she didn't know what could happen or where they may end up, but she was glad to have her friend back. By the time Abby and Sarah made it back to where the others were waiting, they were gone. Abby and Sarah talked it over and decided to go after them. They thought if they were in the building, that Sergeant Scott and Lieutenant Lee would see them, so they decided to look outside. Sarah wanted to let them know they were only safe if they stayed together. When they couldn't find anyone, they decided to go find Sergeant Scott and Lieutenant Lee to let them know what happened. Sarah could tell something wasn't right by the way they were acting. She told them that the others were gone, and she looked everywhere, and wanted to know if they saw them. Lieutenant Lee told her that they haven't seen anyone, but they were busy with Katie, so if they came by there was a chance they could have missed them. Sarah asked how Katie was doing and they told her that she was still out and they used chains, that if she woke up he thought they would have a better chance and escaping. When they were satisfied that Katie wouldn't be able to break free quickly, they left to find the others. As they were walking out, Sarah told Sergeant Scott she wanted to talk to him about what he saw in the warehouse. Before he could respond to her, Molly came around the corner. With a sigh of relief, Sergeant Scott led them into a run down house. Not knowing what was inside they needed to hid fast, but had no time to check and make sure it was safe. Sarah ran in first and Abby followed, but what the two of them ran into could not be good. Sarah and Abby let out a loud scream that sent Molly hiding. When Sergeant Scott caught up to them he knew that this was it, that they had no way out. He turned around to try and run, but what came around the corner made him fall to his knees. Abby looked and Sarah and whispered what do we do now?

Chapter Five

They were surrounded, there was no way out, but what happen next confused Sergeant Scott. He knew Katie was dangerous because of what she did back at the warehouse. When he saw Katie standing there, he knew he was going to have to tell Sarah about the warehouse. He tried making his was to Sarah, but Katie was blocking his way,but all the sudden everyone fell, like they did when it all started. Confused, Sarah looked at Sergeant Scott, she could tell something wasn't right so she asked him what was happening. He told her, he didn't know, but she just couldn't hear that same answer anymore. She told him that she deserved to know everything that was going on and that she wasn't going to take I don't know as an answer anymore. He told her that when they got some place safe he would tell her everything he knew. She was happy about it ,but told him fine , but she wanted to know everything not just what he thought she should know. They knew they needed to find somewhere to go , and didn't know how long they had until Katie and they others would wake up. So they headed out to find a safe place they could call home. After walking all night they came up to a church, it was surrounded by a fence. They didn't know what would be behind the fence, but it looked safe, so Sergeant Scott went first to make sure it was safe. After searching, he saw it was empty and went to get the others. He told them it was safe, that it was completely abandon. He told them, that would be a good spot, that they could get some rest. Lieutenant Lee said that he would be on look out, and the rest could take a look around to see what there was for them to survive. They found food in a small kitchen just off the church. Abby decided she would cook them something to eat, while Sergeant Scott and Sarah continued to look around. Sarah couldn't understand why no one was here, everything look perfect. Abby knew this place was to good to be true, there had to be a reason why no one was here. She took out the food to a little table, she told the others that she didn't think they should get to comfortable here, that something felt off. Sarah told her they wouldn't until she knew for sure it was safe. Satisfied, Abby went back to the church where she cleaned up so they could get some rest. They thought someone should always be on watch, so they took turns. It was a quite night , which made

Sarah happy, because she wanted this place to work out. At first light, Sarah decided to hold a meeting to let everyone know she thought that two of them should go out and get supplies and two should stay to watch the place. Abby and Lieutenant Lee decided to stay at the church while Sarah and Sergeant Scott went out for supplies. Hoping it would be an easy day, Sarah was eager to be out. She just knew they would find what they needed, and was hoping they would find the group they lost. Back at the church Abby was walking the fields, when she came across a little house, hoping there was supplies inside she opened the door. When she saw what was in there she let out a loud scream, which sent Lieutenant Lee running. When he got there Abby was on here knees crying. He put his hand on her shoulder and that's when he saw a baby laying on the floor. Not knowing if the baby was infected or not, Abby went to find gloves or something so they could use check the baby out. She found a thermometer so she could see if the baby had a fever. With the baby, there was a bag and a note that read" please keep my baby safe, it is not safe out here for him, I know whoever may find him will take care of him and keep him safe. Please don't judge me for this." In the bag there was everything they needed to take care of him. After they checked him and he seemed to be okay, Abby couldn't help but cry. She was so happy that this sweet little boy was okay and not infected, because she wouldn't have the heart to hurt him. She changed him and feed him, then she found a basket and laid him down so he could sleep. Just before the sun went down Sarah and Sergeant Scott returned. They had supplies and they had found the group they lost. Sarah went in first happy to let Abby and Lieutenant Lee know what they found, but when she saw they baby laying in the basket, she was confused and worried as to where he came from. She asked Abby where he came from, so Abby took her to the house where they found him. Sarah didn't like this at all, she knew this house was not safe. She asked Abby if they checked the house, Abby told her no ,that they just took the baby and didn't even think to check it. Sarah told her it was getting dark, that they should head back and worry about it in the morning. When they got back to the church Abby got the others feed and fixed them a place to rest, while Sarah talked to Sergeant Scott. She told him about the house Abby took her to where they found the baby. She told him that the house just didn't feel right and she thought someone should check it out, but wanted to wait until morning so they could see. He told her that he didn't know if they should wait or if they should go now. She told him she didn't want anyone to know , that they should stand guard together while everyone slept. When he agreed Sarah went and told everyone that her and Sergeant Scott was standing guard tonight and that they should get some rest. After everyone was asleep Sarah and Sergeant Scott decided to go to the house to check it out. They didn't want any surprises during the night. When they got to the house Abby was standing at the door, she told them she needed to know if there was answers to why this baby was left. Sarah talked to her for awhile and finally got her to go back to the church to take care of the baby. After she was out of sight Sarah and Sergeant Scott went inside, Right away they knew something was not right. As they went room to room through the house they found a small room at the back of the house. Startled by a loud noise , Sarah let out a scream. They did not know what they were about to walk in to. Sarah was hoping no one heard her scream and they decided to open the door. Sergeant Scott, gun drawn, opened the door. Nothing could ever make them ready for what they saw. Back at the church Lieutenant Lee heard a gun shot and went running. When he got to the house he couldn't believe what he saw.

Chapter Six

Back at the church Abby told the others that she knew something was off and that they should prepare for a fight. Everyone wanted to know what was going on, they thought this was a safe place and that they would finally get some peace. They kept asking where Sarah and Sergeant Scott was, but Abby didn't want to say anything until she knew for sure, so she told them, that they were just double checking to make sure it was a safe place to stay. She knew it had to be something wrong with the little house they found, but she tried to remain calm so the others would not worry. She knew that whatever it was that Sarah and Sergeant Scott were the best two to handle anything, and she knew they would keep them safe, or she hoped. Meanwhile back at the house Lieutenant Lee was trying to figure out why he heard gun shoots, he thought they were safe and didn't know how much more he could take. Sarah told him everything and asked him if he would go back to the church to make sure things didn't get out of hand. After he left, Sarah asked Sergeant Scott knew what they needed to do? He told her that he would take care of everything that she didn't need to worry. If it was only that easy, here she was looking at a crime scene. There was a woman , a man and two small children, the woman was still alive when they opened the door , but charged at them and that was why Sergeant Scott shot her. Sarah told Sergeant Scott that must have been the baby's family, that the mother must have known, so she locked them up in the small room. Sarah headed back to the church to tell Abby that she thought they found the baby's family. Abby grabbed the baby and headed to the house, but before she could make it past Sarah, she had to hear more. Sarah told her there was more, she told her she thought the mother must have known she was sick, so she locked herself up in the room hoping someone would find her baby and keep him safe. She told her that there was a man and two small children in there with her and none of them had a chance. Abby hit her knees, she began screaming and demanded to see the house for herself, but Sarah told her it would be better if she didn't see it, that if she did she could never forget. Abby wanted to know what they were going to do, if they needed to move on to a knew place or if they could make it here. Sarah told her for now they were staying, that Sergeant Scott was taken care of it. Abby carried the baby to Carrie, one of the ladies that has been with them since the lab, and asked her to look after him. Carrie smiled and that's when Abby told Sarah she had to see the house and took off. She had to make sure for herself that this place was safe, because she knew Sarah wasn't telling her everything. When Abby made it to the house Sergeant Scott was nowhere to be found, when she opened the door to the room she saw why Sarah had told her to stay away. It looked like they were all tortured before they fully were infected, and she thought somebody else locked them in, and left the baby because he showed no signs of being infected. Abby couldn't understand how someone could be so cruel, what kind of person could do this kind of thing. She headed back to the church, she wanted answers.When she saw Sarah she was furious, but remained calm for everyone else. She told her she went to the house and saw everything, but that Sergeant Scott was gone. Sarah took Abby outside, she told her to calm down that he had to be around somewhere. Abby told her she knew the truth about the little boy's family, and she could tell that someone else lock them up, not the mother. She told her that the family was used by someone to try and get answers, but didn't think the person that did this was gone. Sarah asked why she thought the person would still be here, Abby told her that she just felt like he or she would be looking for more people to give them answers. She told her she didn't feel safe here and she wanted to know where Sergeant Scott was at. Sarah told her that when she left the house, he was there to take care of the family so no one saw what went on there. She told her if he was not at the house she didn't know where he would be. They both agreed to keep this quiet so no one would freak out and leave. She promised Abby, that she would figure out what happen to Sergeant Scott and that she would keep everyone safe.They both headed back in the church t check on everyone and to fix something to eat. They agreed to keep an eye out in case Sergeant Scott was hurt or infected. Abby stayed inside the church while Sarah was outside making sure they were safe. Abby didn't want anyone to think something was going on, so she searched and found some old games for everyone to play, she was hoping it would take her mind off of what was going on. While playing games, Andrew one of Sarah's researchers remembered Abby saying they may need to prepare for a fight. So he asked Abby what was going on out there? Abby looked at him confused and said " what are you talking about'? He told her he remembered her saying something about preparing for a fight and he wanted to know what was going on. Feeling pressured Abby ran out of the church in tears. Sarah saw her and thought Sergeant Scott showed up so she went after her. When Sarah caught up to her she saw her fall to her knees. Abby heard someone walking up behind her and turned around. Sarah could see something was clearly wrong . What happen? Sarah asked her. Abby told her when she left the house she told everyone that she felt something was off and that they may need to prepare for a fight. Before Sarah could respond Lieutenant Lee came running at them in panic. What is it now? Sarah asked. He told her that they needed to come back to the church, that Sergeant Scott just ran into the church, and took the baby and Carrie. Sarah asked if he said anything? Lieutenant Lee told her, that he told him, that if we wanted them back I needed to tell them the truth. Sarah looked at him and said what truth. Before Lieutenant Lee could answer he collapsed to the ground. Now what is going on, what don't I know, she thought. She knew what needed to be done, so she went in the church, no one was there. They must have already left she thought. So she went back and got Abby, looks like we are on our own.

Chapter Seven

Abby was scared, but she knew her and Sarah could no longer stay here. She was also afraid the two of them could not make it on their own, that they would need to find new shelter fast. Sarah searched the church, to see if they could find a few things to take with them, but the church was empty. She went and told Abby they would have to make this trip without any supplies, but they could look for some along the way. Abby didn't like anything about this trip, not having supplies and it just being the two of them, it just didn't seem right. Abby told her that she didn't like this at all, but agreed its what needed to be done, so they headed out hoping to find something fast. Both were hoping they would run into their group, so they would have more people to help. They headed out preparing for the worst, but not even a mile down the road they saw Carrie. They didn't know if she was infected or hurt, they could see her holding something. They took caution as they past, but when Carrie spoke to them, they knew she was alright and saw she was holding something. Sarah asked her what happened to Sergeant Scott. Carrie told her that after he grabbed them, that Katie came charging at them and that she knocked Sergeant Scott down a hill, and went down with him, so that was when she ran. Sarah asked if anyone else had came by. Carrie told her that, her and Abby were the only ones she had seen, and that she was about to give up hope of being found. Abby told Carrie that everyone had left not long after Sergeant Scott took them, but the must have headed the other way back to the lab. Why would they go back there, we all know what is back there, it just doesn't make any sense. Sarah told them not to panic, that they would head away from town, that she didn't want to risk anyone getting hurt. She told them she hopes the others would realize the are headed the wrong way and turn around before it was to late. She helped Carrie to her feet, and Abby carried the baby to give Carrie a break, and they headed out to find shelter. They walked about four miles when they came up on a small store, the three ladies decided to go in and check it out. They were tired and hungry and running out of hope for finding food. When they entered the store Sarah smiled, because it seemed to be untouched. She was happy because things seemed to be looking good for them for once. They each grabbed a shopping cart and filled it with things they needed to survive , Abby found a stroller for the baby so they could take a break from carrying him. Happy with what they found they headed out to find shelter , even if it was just for the night. They walked about four more miles when the came across a house. There was no fence which made Abby nervous, she thought it would be to easy for someone to just walk up on them. She told Sarah she didn't think this house was to safe, but Sarah told her it would have to work, they needed to rest and they could block the door and windows. Sarah told them to wait outside to let her look and make sure no one was inside, after finding nothing she went out and got them. She told them the house had power and hot water, that they should shower and eat then get some rest. After they showered, while Abby fixed them some food ,Sarah made sure the house was safe and Carrie fed the baby. While they were all sleeping, Sarah heard a loud noise that woke the baby. She tried to calm him, but he wouldn't stop screaming. Abby woke up asking what was wrong with the baby, Sarah told her she heard something upstairs. She told Abby to watch the baby while she went to check it out. When she got to the top of the stairs she heard the noise coming from the master room. She opened the door and that was when she saw Molly trying to break in the window. She ran down to get Abby and Carrie, she told them to get the baby they had to go now. Without any questions they picked up the baby, but before they got to the door they heard the window upstairs break. Sarah told them Molly was upstairs to run and run now. They ran as fast as they could, but when they turned the corner Sergeant Scott was standing there just staring at them. They turned to head back to the house, but Molly busted out of the house. The woods Abby screamed, head to the woods. They ran in the woods, they came up on a cabin, they were tired and couldn't run anymore. They got closer to the cabin, when they saw a shadow of someone or something. They wanted to head back out of the woods, but they still heard them following them. They turned back to the cabin, the shadow was gone. They ran up to the cabin, but what they saw brought them to their knees. Sarah cried out, God help us.

Chapter Eight

Sarah didn't know what was going to happen to them, she was afraid that the men that took them could be infected. Not knowing if Abby, Carrie or the baby was okay she started to panic. They kept her in a small room, with nothing but a mattress. She decided she needed to have a look around, but when she got to the door she was surprised to find the door unlocked. When she walked out the door there were two men, one on each side of the door. She started to turn and run away, when one of them grabbed her. She started to scream, because she was scared of what he would do. When he finally got her to calm down , he told her that they saw her fall while they were out searching for fire wood. He went on to tell her that she was screaming and begging for help, that she said a man was after her and her friends. "My friends" she gasped, where are they? He told her when she was ready he would take her to them. Sarah followed him down a long hallway. He oped a door where she saw Abby and the baby, but no Carrie. When Abby saw Sarah she ran up to her, with tears rolling down her face. Sarah asked her where they were at and what happened to Carrie. Abby told her that four men came out of nowhere,they found her on the ground and helped them up in their truck. Abby went on and told her that she had passed out when they brought them here. Why did I pass out? She asked Abby. She told her that they all were so exhausted from running and that she was just worse off. She told her that her and Carrie, along with the baby were checked out right away, but they put her in a room until she woke up. Abby told the men that they should take them to Carrie,so Sarah could see that everyone was okay. Carrie was out in a garden picking fruit and vegetables for them to eat. When she saw Sarah, she dropped everything, she was so happy to see that her friend was alright. After Sarah saw her friends were fine, she asked if she could have a look around, so she could see for herself that they were safe. The men told her to knock herself out, that they would be up at the house in the den if she needed them. She smiled and went to have a look, She saw they had plenty of food to last them awhile, but she was a little concerned that there was no fence around the property. Confused about the fence she went to have a look inside the house. Nothing seemed off until she came up on a door leading down to the basement. When she tried to open the door it was locked. Not happy about no fence to keep anyone or anything out, then finding a locked basement door just didn't seem right, so she went to find the men. They asked if she was satisfied with everything and she told them that there was a few things that she was concerned about. They told her there was nothing to be concerned about, but to go on and tell them what she was worried about. First she told them she was worried that there was no fence to keep anyone or anything out. They told her they saw no need for a fence, that only people that had something to hide would have a fence. Confused she asked, If you have nothing to hide , then why is the door to the basement locked? The basement is no concern of yours, one of the men shouted. Sarah ran off to find Abby and Carrie. She told them that something wasn't right about this place, that it was time for them to go. Just as she finished telling them what she found, a man came up behind her. He grabbed her arms and said, now its time for you to see the basement. Sarah fought them off and took off running, she was hoping the other two would follow her. She was halfway in the woods, when she noticed Abby was with her, but Carrie and the baby was not. When Abby noticed Carrie did not follow and had the baby she began to panic. What do we do now,Sarah? Abby shouted. We just have to stay calm, she told her. Calm,what has staying calm gotten us? Sarah told her that when it got dark they would go back and get them, but if she didn't stay calm they would not make it. Meanwhile, back at the cabin, two men locked Carrie, along with the baby in a room, and placed two men to stand guard. They knew Abby and Sarah would come back for Carrie and the baby, so they had to be ready when they did. Carrie wasn't worried, because she knew Sarah would come back for them. Sarah and Abby searched the woods to find anything they could use as weapons. After not finding much, Sarah told Abby they would have to use what the had. Just as it was getting dark the two of them headed to get Carrie and the baby. The only thing Abby was concerned about was that sweet little boy. Sarah grabbed her by the arm and told her they needed to head back to the woods. No, Abby screamed, we can't go back they need us. Can't you see this is a trap? Sarah asked. They need us,Sarah, how can you just leave them like that? Abby told her, they had to keep going, they have came this far, it wouldn't make sense to go back. Just then, two men came out of the darkness and grabbed them both. Sarah tried to fight them off, but they just laughed. Looks like we got ourselves a fighter here, one of them said. He just smiled at

Sarah and said, not this time sweetheart, we got plans for the both of you. He knocked Sarah out, which made Abby scream. They just looked at Abby and said, don't worry that pretty little head, we got plans for you too. Should we knock her out too, one asked. No not this one, we will just tie her up and she can see what we do to her friends. That's when Abby knew she needed to fight, so she could save her friends. All the men could do was laugh, they even let Abby go so they could chase after her. This went on for about two hours, when Abby knew this could be it. She didn't want it to end like this, she knew if they got her into the basement it was the end. She started thinking about what Sarah would do if she wasn't passed out. She knew Sarah was stronger than she was, but she also knew Sarah would never give up. They let Abby go one last time, she knew she had to make this one count, so instead of running out to the woods, she took off after Carrie. That was when one of the men grabbed her by her hair and dragged her to the basement, and that was when everything went dark.

Chapter Nine

Carrie was watch out the small window in her room, when she saw two men taking Sarah and Abby into the house. She still didn't panic, because she knew Sarah would not give up. Sarah and Abby were separated, neither knew what was going on. Like Carrie, Abby didn't panic, she knew Sarah would find a way out.When Sarah came to, she started to look around, she knew she had to find a way out to help her friends. She saw no door, no windows, but she knew there had to be a way out, because they got her in somehow. She could hear footsteps above, and she could hear people talking. That is when she knew they had her underground somehow, then she heard Abby screaming her name, she could hear the fear, that is when she knew she would have to use everything she had to get free. She tried to be as quiet as she could, she didn't want the men to hear, because she didn't know what they might do if they heard she was awake. After a few hours of trying to break free, she began to think she may never get out. That is when she knew she had to try one last thing. She made as much noise as she could, hoping someone would come to see what was going on, that way she would know how they got in and then she could get out. After trying that for a couple hours, she was about to give up, but then she heard footsteps. The footsteps stopped right above her head. A man opened a small door, and dropped down a small rope ladder, then he climbed down. He slapped her and told her, that if she wanted to live, if she wanted her friends to live,, that she needed to shut up and behave like a good little girl should. She said nothing, she just laid on the floor, she wasn't scared, she just knew it would be a little harder to get out. Right before the man shut the door she hear Carrie yell out her name., than it all went quiet. This was when Sarah started to panic, she stopped sleeping , she knew she would have to do all she could to get them to come back. She knew her friends needed her, but no matter how loud she was no one came. It seemed like months went by, she didn't know if anyone was still alive or not. She was so tired, she hasn't had food since she has been in this place, she didn't know how much longer she could go. She began think okay this is it, they will come today and I can help my friends, but nothing. By this time she was ready to give up, but then she heard a familiar voice yelling her name. It was Sergeant Scott, she could hear panic in his voice. Then she heard gun fire and yelling, she didn't know what was going on, for once she was happy to be in this hole. She heard footsteps, she knew they were getting closer. The footsteps stopped, then she heard Sergeant Scott yell, Sarah are you down there. She yelled back, but wasn't sure if he could hear her or not. When the door opened and she saw him, she started to cry. He was covered in blood, then she saw Abby come up behind him. He told Sarah not to worry, he would get her out. He let down the rope ladder and climbed down after her. She could barely stand much less climb up a ladder. He told her he would help her out, so he pushed her up the ladder as he climbed up after her. Sarah laid on the ground crying, she was so happy to be out of that hole and see Abby. She was so weak, she told Abby and Sergeant Scott that they should have left her down there, that she was no good to them now. Abby laid down beside her and told her that without her she would have been died a long time ago. She went on to tell her ,that she is her best friend and that she would not leave here without her. Sarah told her she had no more strength to go on. She told Abby that this was it for her, that she just needed to go with Sergeant Scott and live her life. Abby talked to her for awhile, trying to get her to understand she would not do this without her. After watching them for awhile, he could tell how tired Sarah was, but he could also see she wasn't ready to give up. Sergeant Scott went and helped her up, he told her he was going to help her get out of this place alive. He told her that she needed to prepare herself for what she was about to see, that he had to fight his way to get to her. She told him she understood, that she had to do some things she wasn't proud of.He helped her to her feet, and told her if at anytime she need a break to let him know and they would stop. She nodded and they headed off, he looked back at Abby and told her she should be prepared, he had to fight to get to her also. He told her if things got to hard to see to let him know and they could rest. After he made sure both girls were okay , they took off. He was hoping they could find a safe place for the night that wasn't to far, because he knew Sarah couldn't make it far. Both girls couldn't understand why he didn't get Carrie and the baby first. Before they turned the corner where Carrie and the baby was kept he told them, he had to tell them something, but it was going to be hard to take. They both looked at him confused and both broke down, please just tell us what is going on. He looked at both of them and said I just want you to know that I was planning on stopping here first before I got the both of you, but there were so many of them blocking my path I had to find a way around. He told them they just needed to prepare themselves for what they were about to see, and he was so very sorry he couldn't stop it. Sarah and Abby just hit the ground screaming please don't tell us that, we don't believe you. He looked at the two of them and opened the door and told them if they didn't believe him to take a look for themselves. They walked in to the room and just hit the floor, why did this have to happen?

Chapter Ten

Sergeant Scott told them that he found them like this when he finally made his way to them. He told them that he had it planed to get Carrie and the baby first, but it was blocked, and the way he took, lead him in a circle behind Carrie's room and he saw them laying on the floor. He said when he saw them laying there it set him off, and that's when he just started shooting. He told him that was the moment he knew he had to get to them, before he lost them too. Sarah asked, him why he took Carrie and the baby from the church in the first place? He told Sarah that after he read the note that the mom wrote for the baby, that if everyone found out what had happened to them, he was afraid they would harm the baby and he needed Carrie because she was so good with him. He told her he knew he needed to act fast and get the baby to safety. He told her once he got Carrie and the baby to a safe place he was going to find Sarah and let her know what was going on, but Katie came out of nowhere and that's the last time he saw Carrie or the baby. If it wasn't for you they would both still be alive,Abby shouted. You should have just left things alone,we could have kept everyone alive, Abby yelled. Sergeant Scott broke down, he told her he knew he messed , he knew it was his fault that Carrie and the baby were dead. He told her although he wasn't here when it happened it was because of him they were dead. Sarah told them both it was enough, no one could have known what was going to happen. She told them to get their selves together it was time to move on. They agreed with her, and picked themselves up and got ready to head out. Then out of nowhere they heard a loud cry, Abby looked at Sarah and said, there is noway he could still be alive. Sarah walked back into the room and sure enough that sweet little boy was still alive. How is this possible,Abby asked. Sarah told her it looked like Carrie protected the baby until she took her last breath. Abby looked at the baby and told Sarah it was time they gave him a name. What should we name him? Asked Sarah. Abby told he,r while her and Carrie talked, that Carrie told her that she has always wanted kids and if she was blessed with a son she would name him Toby. Sarah smiled.and told Abby that was a perfect name for him. After they cleaned Toby up, they went out searching for a safe place to rest. Neither one of them knew what was in store for them out there, but there was one thing they knew for sure, and that was they would all work together to keep Toby safe. They walked for hours, and it was getting harder for Sarah to stand, and Toby would not stop crying. We have to find a place to rest and some food, Sarah said. We just have to make it just a little further,Sergeant Scott told them. They walked for another hour or so when they came up on a little town. It looked as though it hasn't been touched with the disease yet and they were hoping they would be welcome there. A woman on her way to work saw them and decided to see if they needed any help. She stopped and asked if they could use a hand ? Sarah told her they have been walking for hours and could use a place to rest, a shower and maybe some clean clothes. They lady told them to get in and she would see if she could find them something. She took them back to her house, and told them they could shower and gave them some clothes. When they were done she gave them some food , she told them they could stay here and she would check on them after she got home from work. Although Sarah was exhausted she felt like there was something off about this place, but she just couldn't figure out what it was. Abby was in the other room with Toby trying to get him to sleep. Along with Sarah, Sergeant Scott thought there was something off about this place. He walked over to Sarah, and told her that something was odd here, and he couldn't figure out why this woman would just pick them up, and take them back to her house, then leave. Sarah agreed,she told him they need to look around to make sure they were safe. So they looked around the house , but all the doors they tried to open were locked. This is strange, Sarah said, why are all the doors locked? They went to try the front door, but it was also locked, but when they tried the back door it was opened. They looked at each other and walked out to the backyard, where they saw six graves. They turned around in panic, to get Abby, but there were four men standing in there way. Wished you didn't do that one of them said. If you would have kept to yourself and stayed with your friend we wouldn't have to do this. Before Sarah could ask what he was talking about, the woman walked in the house. She looked at the men and nodded, then they took Sergeant Scott out back and knocked him on his knees. They told him he only had his friend to think for this. One of the men told him that he should have told his friend that she needed to learn not to go sticking her nose in other peoples business. Sergeant Scott told him that his friend isn't the one burying people in the backyard. Oh we got us one who likes to talk, one of the men said. The woman walked outside and whispered something in one of their ears. Sergeant Scott looked up at them and said, if you would have just stayed away I wouldn't have to do this. Sergeant Scott stood up, but what he saw when he turned around made him freeze in place.

Chapter Eleven

No, No , this can't be, Sergeant Scott kept saying. How can you be here, Sarah, he yelled. Sarah fought her way out to him, but froze when she saw why he was so upset. How can this be, how can she be here? Sarah asked. She look over at Sergeant Scott, she could see how puzzled he was. I don't know what kind of games you are playing here,but it's not funny. She started to cry,she couldn't understand how this could be happening. The men let sergeant Scott go,he ran over to Sarah. She was dead,he told Sarah. Sarah pushed him away,and started walking over to her. She put her hand on her shoulder, and in a soft voice she said, Katie, its me Sarah. Katie just stood there,she didn't say a word. What have you done to her,asked Sarah. She was trying to make sense of all this,but no matter how hard she just couldn't understand how this could be. She knew there was something off about this place, but nothing could prepare her for this. She went back inside the house, she wanted answers to what was going on. Everyone inside just looked at her like she was crazy,but she knew she wasn't crazy, although she started to feel like she was. Sarah kept looking outside,and all she saw was Sergeant Scott, and all he could do is stare at Katie. He started walking over to Katie, but then he paused. He looked back at Sarah, who was looking at the window watching. When she saw someone, or something move past Katie, she began to panic. She could see it was heading straight for Sergeant Scott, that was when she ran outside to help him. When she made it halfway to him,she froze,once again. Sergeant Scott had no idea Sarah had came out to help him. He wanted to get to the door and lock it so everyone inside was safe. He was afraid to turn around,because he didn't know what would happen if he turned to run. He took his time,but move as quickly as he could, walking backwards toward the door. Sarah was still frozen and didn't realize what Sergeant Scott was doing. Then someone grabbed his arm,he screamed and that made Sarah jump. She could see that he was trying to get to the door. Sarah took off running, trying to make it to the door. She wanted to get there before Sergeant Scott, because she was afraid he would lock her out,and she didn't want to be out here with Katie. As she was running she was yelling for Katie to come with her,but Katie just stood there. Sergeant Scott barely made it inside the door,they got the door shut and locked,just in time. At that time Abby came around the corner and asked what was going on. Sarah looked at her all confused,and asked her how she didn't hear all the screaming until now. Abby told her she hasn't heard a thing,so she decided to come and make sure they were okay. Sarah just stared at her,we have been yelling and had people try to kill us,what do you mean you haven't heard anything. Abby ,who was now confused, told her she has been rocking Toby,and must have fallen asleep. She had no idea they were in trouble,or she would have been out here sooner. Sarah hugged her and told her it was okay,that she knew they have had a long journey and she needed rest. Abby told her, this was the first time she felt safe in a long time.Sarah told her they were not safe here,she took Abby to the window,so she could see Katie. Abby just stood there for the longest. Sarah asked Sergeant Scott if he would watch Abby,so she wouldn't go outside. He agreed,and Sarah went to make sure Toby was safe,she was hoping the house was safe enough for now. When she got to Toby,the woman who had helped them,she was sitting there rocking Toby. When Sarah saw her ,the lady looked up at her and said, it has been awhile since we had a baby around here. Sarah stared to panic,she didn't know what this woman was going to do,but the lady walked over and handed her Toby and walked out of the room. Sarah laid him down and she knew it wasn't going to be safe as long as they had company. Sarah went back out to Abby,and Sergeant Scott. Abby looked at her ,tears rolling down her face, and asked how can this be,you told me she was dead. Sarah could see that Abby was confused and hurt. Abby sat crying,she was trying to wrap her head around the fact, that her friend was standing out there. Sarah,Abby cried. I thought she was dead,how can the dead be outside? I know this is all confusing,Sarah told her,but she is gone, we can't help her anymore. Abby screamed, no I won't except that, look at her she is alive. We have to go out and help her,Abby screamed. Sarah grabbed her and told her,you can't go out there,she is gone we need to find our way out of here and protect Toby. Sergeant Scott told them,they would wait until Katie and the others were gone and they would go out the back. Toby let out a loud cry,Abby got up to go get him,but Sarah could see she was in no shape to take care of him. Sarah went and got Toby a bottle,while Abby just sat and cried. Sarah couldn't understand why Abby was so upset about Katie. After she feed and changed Toby, she put him back down so he could sleep. She went back to Abby and Sergeant Scott,where they sat for hours watching Katie. Finally the crowd took off,Abby ran out of the house to where Katie had been standing. She could tell something wasn't right with her. By this time Sarah could see something was wrong,so she ran out to Abby. When she got there,Katie turned around and looked at the two of them and said, You shouldn't have come here,now I have no choice.

Chapter Twelve

Both Sarah and Abby looked at each other wondering what Katie was talking about. Before the two could ask,Katie let out a loud cry. They just stood there,they were to scared to move. Abby started to mumble,but wasn't making sense. Sarah kept asking her, what she was trying to say,but before Abby could mumble another word,Sarah saw Kelly,Adam and Emily.They ran right past them and went right to Katie,who was standing with Carrie. After about an hour the five disappeared inside the house. As soon as Abby and Sarah could move again,they went in looking for Toby.You could hear Abby yelling through the house,you could hear panic in her voice. Abby came back to where Sarah was,she was crying. He's gone,Abby cried,what are we going to do. Sarah was still trying to figure out what was going on around here. Nothing made sense anymore. The house was empty,so Sarah and Abby went out to search the town,but it turned out they were they only people left in this town. Before they decided if they should stay or leave,they had a look around. The stores were still full,they had everything they needed to survive. Sarah was confused,this place was perfect,what could have made everyone leave. They decided to stay,so they got two carts and filled them up with what they would need to survive,they got more than the needed,because they didn't know if it would be safe to come back out,once they found a place or not. After they got the things they needed, they headed out in search of a place to stay. They came across this beautiful house. It was a large two story white house, with a white fence surrounding it's property. This is our house, Sarah told Abby.Before they got comfortable they decided to look around. After they searched the house and property, and they were satisfied,Sarah took the groceries to the kitchen and told Abby she would fix them some food. Abby told her she would figure out where they would be sleeping for the night and if she needed help to yell. Sarah and Abby sat down at the table and ate. While they were eating they discussed what they should do to survive. They wanted this to be the place they called home. After they ate and got things cleaned up, they made sure the house was secured before they went to sleep.The next morning when they woke up everything was so quiet, it was to quiet. Sarah told Abby that she was going to have a look around,because it was just to quiet around here. Abby told her she was going to stay here,but if she ran into trouble to give her a yell. Sarah felt like she would be fine,but she was hoping to run into something, because it was getting boring around here. While she was out she saw Sergeant Scott walking on the side of the road,with something in his arms. When she approached him, that is when she saw it was Toby in his arms. Sergeant Scott looked as though he went through hell to get this little boy back and keep him safe. She knew Abby would be so happy that Toby was alive. She told Sergeant Scott she was going to take him back to where her and Abby were staying. When Sarah got back,she didn't have to yell at Abby,because she was sitting on the porch with a book. When Abby saw Sarah with Sergeant Scott, she ran out to help her, that is when she saw he had Toby with him.Sarah told her to take them both inside, that she was going down to the store to get some things for Toby.Abby told Sergeant Scott to lay on the couch, that he must be tired and hungry. When she tried to take Toby from him,he snapped at her.She left him alone and went into the kitchen,where she broke down. By this time Sarah was coming back from the store,she was excited to tell Abby the news she had,but was afraid how she would take it.Abby could tell that Sarah had something on her mind, but she didn't care. When Sarah tried to talk to her,she cut her off.Why don't you just take this out to Sergeant Scott because I can't do this anymore,Abby shouted. Sarah took out the food and let Abby cool off before she tried to talk to her again. Sarah walked out on the porch,where she saw Abby. Abby apologized, and asked Sarah what was going on now. Sarah smiled,she told her while she was out getting Toby somethings she came across a lab. She went on to tell her that she went in and saw test results for what was going on with Katie and everyone infected. What does this mean for us now,Abby asked. Sarah told her that now she had the answers she needed ,and they could head back home.When do you want us to leave,Abby asked. Sarah told her they would give Sergeant Scott a few days to rest,then they would go. When Sergeant Scott was back to his self ,Sarah told him what she had found and what their plan was. They gathered up their things and headed back home so they could fix the wrong Katie caused.